SEA JELLIES

Rebecca Woodbury, Ph.D., M.Ed.

Gravitas Publications Inc.

Sea Jellies

Illustrations: Janet Moneymaker

Sea Jellies
ISBN 978-1-950415-58-8

Published by Gravitas Publications Inc.
Imprint: Real Science-4-Kids
www.gravitaspublications.com
www.realscience4kids.com

RS4K

Photo credits: Cover by markrhiggins, AdobeStock; Title Page by Derek Keats, CC BY SA 2.0; Page 7: Left, By SeanWonPhotography, AdobeStock; Right, By roza27, AdobeStock; Page 9: 1. W. Carter, Public Domain; 2. Ed Bierman from Redwood City, USA, CC BY SA 2.0; 3. Adrian (User-Intandem) at en.wikipedia, CC BY SA 3.0; 4. Alexander Vasenin, CC BY SA 3.0; 5. Seascapeza, CC BY SA 3.0; Page 21: 1. NOAA; 2. Derek Keats, CC BY SA; 3. Fred Hsu, CC BY SA 2.0; 4. Orest, CC BY SA 2.0; 5. Papa Lima Whiskey at English Wikipedia, CC BY SA 3.0

Have you ever seen a sea jelly?

Does it go on toast?

- 3 -

Sea jellies are sometimes called jellyfish, but they are not true fish.

Sea jellies are soft-bodied animals that live in oceans.

Sea jellies have different

shapes and sizes.

The body of a sea jelly is called the **bell.**

A sea jelly swims by moving water in and out of its bell.

Ding Dong!

Not that kind of bell!

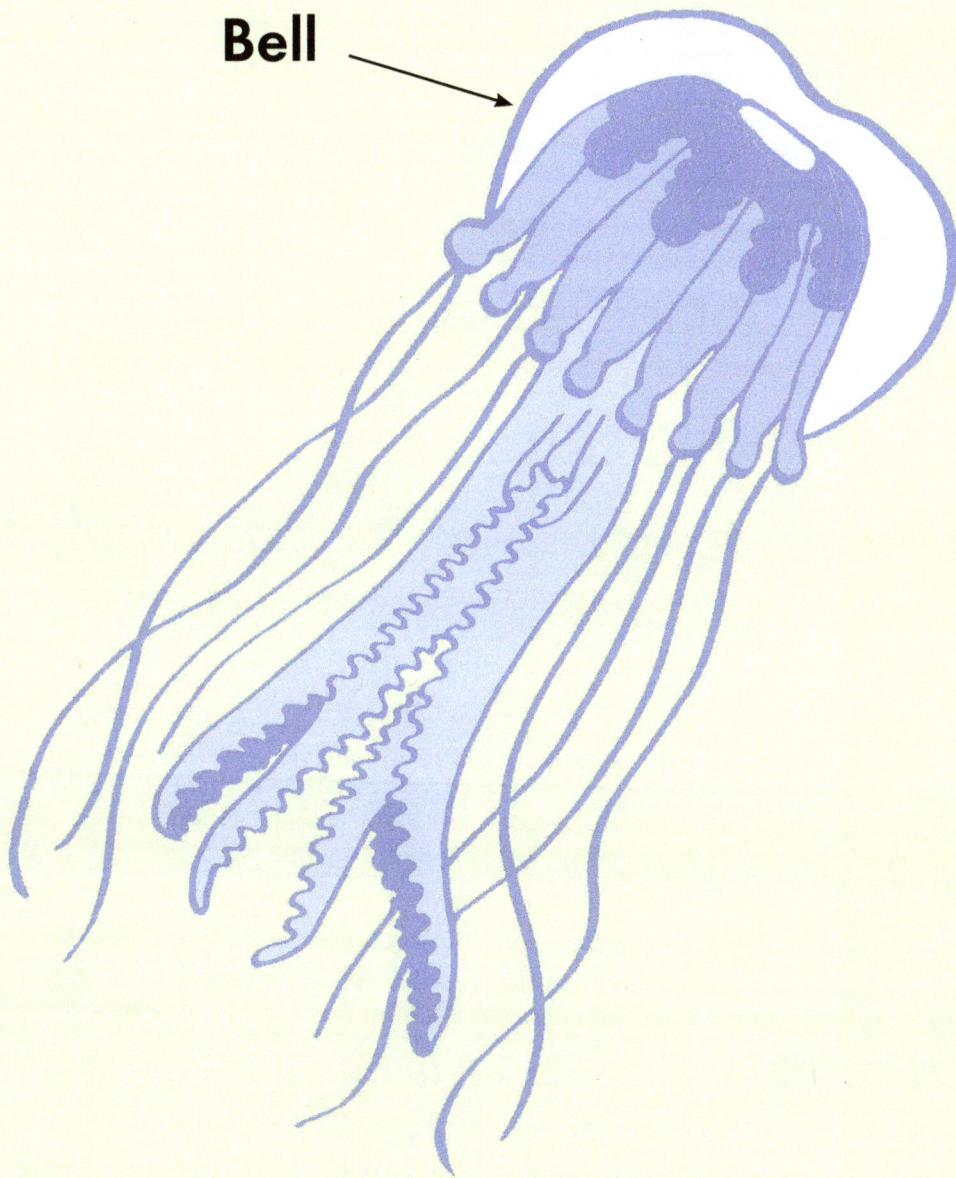

Bell

Some sea jellies have

tentacles.

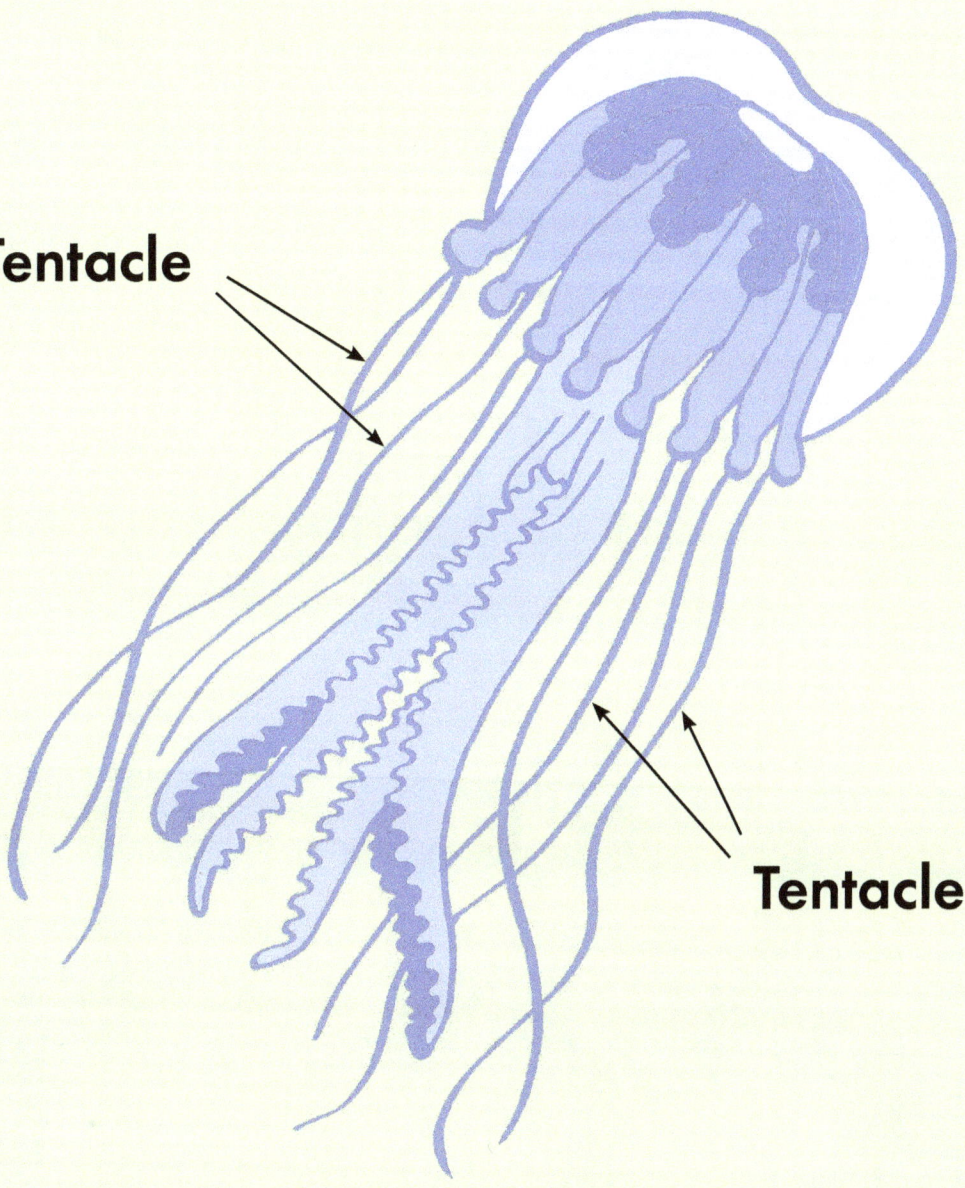

Tentacle

Tentacle

Tentacles help sea jellies catch food.

Sea jellies also have

oral arms.

What does
oral mean?

Having to do
with the mouth.

- 16 -

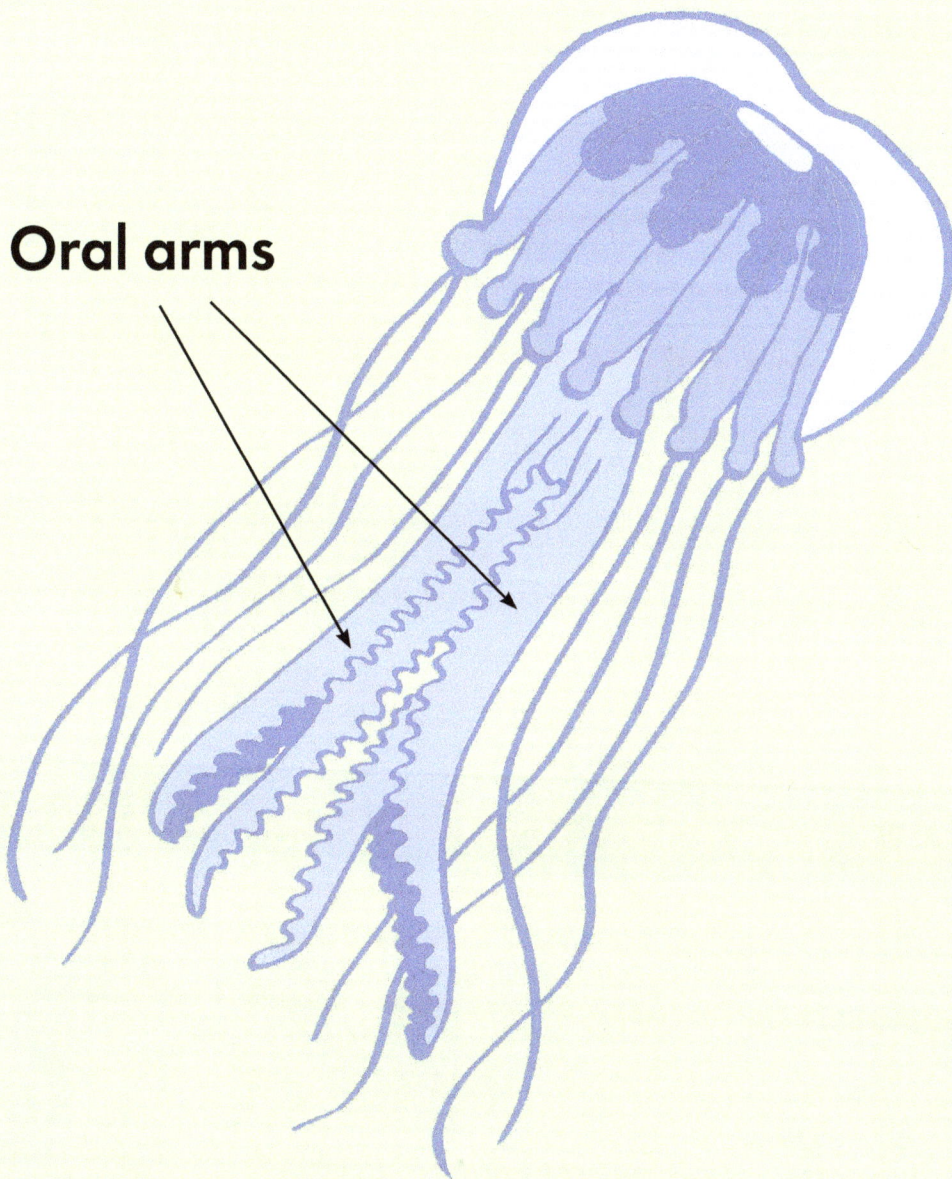

Oral arms

Oral arms move food into

the mouth of the sea jelly.

Do you think
they eat cheese?

No!

Many sea jellies have beautiful shapes and colors.

They are found in oceans all over the world.

How to say science words

bell   (BEL)

jellyfish   (JE-lee-fish)

ocean   (OH-shuhn)

oral arm   (AW-ruhl ARM)

science   (SIY-uhns)

sea jelly   (SEE JE-lee)

tentacle   (TEN-tih-kuhl)

www.ingramcontent.com/pod-product-compliance
Lightning Source LLC
Chambersburg PA
CBHW040152200326
41520CB00028B/7575